Andrew Davies

Beekeeping

Inspiration and practical
advice for beginners

National Trust

First published in the United Kingdom in 2007 by Collins & Brown.

This edition first published in the United Kingdom in 2017 by
National Trust Books
1 Gower Street
London, WC1E 6HD

An imprint of Pavilion Books Company Ltd

ISBN: 978-1-90988-198-3

A CIP catalogue for this book is available from the British Library.

20 19 18 17
10 9 8 7 6 5 4 3 2 1

Reproduction by Spectrum Colour Ltd, UK
Printed and bound by 1010 Printing International Ltd, China

This book can be ordered direct from the publisher at
www.pavilionbooks.com, or try your local bookshop.
Also available at National Trust shops and
www.nationaltrustbooks.co.uk.

BEEKEEPING

WHY KEEP BEES?

The one great thing about keeping bees is that they never fail to surprise you. That's not to say they are unpredictable creatures. There is a logic and efficiency in most of the things they do. But even the most experienced beekeeper gets taken aback by something that will happen over the course of a beekeeping season. Brother Adam, who spent a lifetime trying to perfect his Buckfast strain of bees – a combination of docility and productivity – and who died at the venerable age of 98, was still learning after 70 years as a beekeeper.

THIS IS THE JOY OF KEEPING BEES. Apart from the fact that you can get a crop of honey that is superior to almost anything you will find in a supermarket, there is the thrill of watching them at work and seeing how they react to the changing year.

In an age that is conscious of its environmental responsibilities and "food miles" there is nothing so carbon responsible as bringing in sugar from the garden. Bees make use of a resource that would otherwise go to waste.

Honey is a combination of plant or insect-produced (honeydew) sugars with tiny amounts of pollen and other trace elements. Whereas if you started a vineyard in your back garden you couldn't hope to produce grapes or wine that was on a par with the great vineyards of France, with honey that's not the case.

The vast majority of honey on sale in the UK comes from overseas sources such as China and South America; these are blended and heat-treated to give a consistent but bland final product, as the subtle volatile scents and tastes are lost in the processing.

There is nothing, simply nothing like the exquisite smell and taste of apple blossom honey, and a novice beekeeper can produce this almost as easily as someone with years and years of experience. In fact it would be extremely difficult not to produce better honey than is on sale at the supermarket.

There is also a further analogy between the wine grower's *terroir* and the beekeeper's patch. The honey that is produced in an apiary is an expression of the flowers in a three-mile radius of the hives, combined with the weather over a season. An experienced beekeeper may be able to get his bees to work harder, but he can't easily change the flowers his bees work. So a novice starting up can produce a fabulously flavoursome crop in the first season.

When people at a party find out you are a beekeeper it will unlock a torrent of questions they have always wanted to ask, but never found someone to answer. There is an immense general fascination with bees – in particular from postmen when they are told that the small, hard-backed envelope they

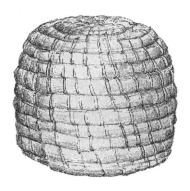

A traditional dome-shaped straw skep

are delivering to you contains a queen bee sent by post with eight attendant workers. Part of this fascination is because bees are still a mystery. We can guess why bees act in certain ways, but we can't be sure. Though some of the secrets of the waggle dance have been revealed, there are many areas of group decision making that have defied scientific explanation.

There are so many good reasons to keep bees. It is profoundly friendly to the environment because it helps the pollination of fruit trees and bushes as well as a myriad of garden flowers. This is why a delayed spring will often produce a more bountiful autumn – because there are more bees around to pollinate everything.

Honey is the perfect gift to friends. Your honey is a unique blend of the nectar gathered from flowers and trees close to your hives throughout the season and it provides you with a product that is well received by friends and neighbours, yet each year is always slightly different.

Beekeeping is a craft that pays you back, or could even provide you with a small income. It is hard not to be inspired by the bees' sheer hard work and persistence and their resilience in the face of blunders by the beekeeper (you knock a hole out of the honeycomb and they go and repair it). It is quite a responsibility looking after a community of up to 30,000 insects and their potential impact on those nearby. But it is a challenge that brings enormous rewards.

This book is not intended to be a textbook for beekeepers. What it will do is give you an outline of what is involved in becoming a beekeeper and what demands it will place on your time and your resources.

THE BEST TIME TO START

There is no perfect time to take up the subject, but winter to early spring is as good as any. You can read up on the subject, contact your local beekeeping group and perhaps enroll on a beekeeping theory course. The majority of these are held from January through to April, though some will be held a little later in the year. Most of the county beekeeping associations will run a beginner's practical course between April & August and getting new recruits in is regarded as vital for the health of the hobby. Through contact with local beekeepers you will be able to see them at work, and most importantly, find out who has some bees or spare equipment for sale.

THE COST

Beekeeping is one of the few hobbies that actually pays you back. Anglers don't calculate the price of fish they catch and balance it with their expenditure on equipment, but beekeepers often do. When starting out in beekeeping there are some things that you should buy new, like a bee veil or suit, gloves, scraper and smoker, and there are some things you can get away with secondhand – particularly hives.

Beekeeping has exploded in popularity in recent years, and so too have the prices. So it's only when you have started keeping bees for a little while that you will be able to work out what is most important to you and then is the time to buy new. There is nothing to match the powerful smell of the red cedar used in the best hives and a visit to a beekeeping shop can be a huge temptation.

Again, the local and county beekeeping associations often have sale days or auctions where you can pick up equipment at a third of the price in the catalogues. In 2016 you could buy a four frame nucleus (mini hive) of bees from Thornes for £260 and a complete but unassembled National hive, along with a beginner's kit of veil, smoker, hive tool, introductory book, record book and feeder for £390.

Though prices vary from region to region you should still be able to pick up a secondhand National hive, complete with bees, for around £100 to £200+ at auction. Or maybe buy a hive from a member of your local association who is cutting down. In all cases you will need assurances that the equipment/bees are disease free.

The great benefit of buying a full working hive is that you will get a good crop at the end of your first summer, while a nucleus will spend the summer building up its numbers and drawing out honeycomb from foundation.

WHAT ABOUT STINGS?

The biggest reason that puts people off beekeeping is the fear of being stung.

One of the great myths about bees is that they are naturally aggressive. Unlike malicious wasps, a bee will only sting when it feels under threat. If a bee lands on a person away from the hive then it will have done so accidentally – it has mistaken you for a flower or shrub or it is simply tired and gathering itself. There is little risk of being stung. When someone comes too close to the front of a hive, though, or knocks into it (bees dislike vibration), then they are at risk of being stung.

Bees have hooks on the end of their legs that can get entangled in hair and in clothes. If they land and then cannot get away, they can panic and sting. It's a reflex reaction. That's why the best reaction to a bee landing on you is to watch it. Provided its legs are not caught in the material of your clothes it will fly away. Rapid movement is most likely to panic it.

A sting is a tragedy for both stinger and stung. What bees don't realise is that when they insert their barbed sting into human skin, the barbs make it impossible for them to remove it from the tough epidermis. And so the sting, along with the venom sac, is wrenched out of the bee's abdomen, killing the bee. When protecting their hives against invasive wasps, bees can usually remove their sting from the chitinous bodies of wasps; not so with other animals.

When trying to establish the culprit of an insect sting, if no sting is left behind then it is likely to have been a wasp that was responsible. Wasps have barbless stings and can sting and sting again.

If stung by a bee and the venom sac is left behind, the best technique to remove it is to scrape it away. Grasping it between two fingers simply pumps the venom from the sac into the skin.

In practice it is impossible to keep bees and never be stung, but with proper precautions, such as tough gloves and a good bee suit, you can keep three hives and get away with only two or three stings in a season. At these rare times it's always best to comfort yourself with the old wives' tale that bee stings help ward off arthritis.

LOCATION, LOCATION, LOCATION

The most important thing to consider when taking up beekeeping is where you are going to place the hives. As the great beekeeping author Ted Hooper says, you can keep bees just about anywhere, but some places are trickier than others.

People keep hives on the rooftops of their London homes, which is fine until they start to swarm…

Providing you keep docile bees, locate them shrewdly, inspect them regularly and at sensible times, there is no way your neighbours will even be able to guess that you keep bees, other than seeing you dressed up in the garden in your bee suit, smoker in hand.

Can neighbours complain, even if you have taken the most stringent of precautions? It is an acceptable standard to have two hives in a suburban garden; any more and you might have to prove you are not posing a hazard to your neighbours. As a beginner you are actually in the weakest position to keep hives in a small garden without problems, as you are more likely to upset your bees with your inexperienced manipulations and not recognise the early signs of such problems.

The advantage of keeping bees in the countryside is that there is more room, giving more space to expand and less worry about the consequences. As we will see later in the book, it is never a good idea to have a single hive because it gives the beekeeper no room to manoeuvre should something happen to their solitary queen and their solitary colony; keeping more than one hive also allows you to compare and contrast the relative differences between your colonies.

Though bees are not naturally aggressive, there will always be a certain amount of disturbance when you open up the hives and start moving frames about. The best time to inspect is on warm, sunny afternoons, which is exactly the time that people are out having barbecues. Be sympathetic to your neighbours' feelings as they are unlikely to be as enthused with your new hobby as you are.

One way of keeping bees out of your own and surrounding gardens is to put a barrier directly in front of the hives, so that they are forced into the air on exiting the hive. This means that the bees' approach is not skimming over your neighbour's fence, but high in the air and much less noticeable.

Another good way of keeping bees out of surrounding gardens is to give them a water source close by. From spring onwards bees will collect water and they will soon track down the nearest source. So if your neighbour has a pond, that's where they'll go and keep going to, so get your local water source set up early in the season before they visit your neighbour's in preference.

In an ideal world hives should point east or southeast to catch the early rays of the sun. This also stops a prevailing southwesterly wind blowing straight in through the entrance. Hives should be at least half a metre apart to reduce bees "drifting" from one hive to another and there should be a source of water close by as this will prevent them wasting energy in transit. Bees use more water in summer, as they need it to help cool the hive.

FORAGE

Bees collect four things: pollen for protein, nectar for energy and conversion to honey, water and propolis from leaf buds to use as glue around the hive. They will journey up to three miles (in some extreme cases five) to gather what they need.

Early season sources of pollen are vital for bees because it allows the colony to grow rapidly in the spring as the protein is in demand from the ever-increasing number of larvae. The main sources of early protein are daffodils and crocuses, so bees on the urban fringe have a distinct advantage over their rural counterparts.

Good sources of nectar are the early blossoms of cherry and apple trees, and shrubs such as viburnum. Later on, tree flowers like horse chestnut and shrubs such as ceanothus provide rich nectar sources, along with agricultural flowers such as field beans or oilseed rape.

The conventional way of extracting normal multi-flowered honey is to put the frames holding the honeycomb upright into a centrifuge and rotate it rapidly. The honey is flung outwards onto the wall of the circular extractor and slides down the wall to be collected at the bottom. The honeycomb that was used to store it can then be put back to work. But special attention needs to be paid to the following nectar sources:

OILSEED RAPE: Honey produced from oilseed rape (OSR) crystallises rapidly and though it is harvested in liquid form, once stored in honeycomb it soon hardens, to such an extent that the bees themselves cannot get it out. This means that beekeepers must remove it from the hive early in the season before it ruins their honeycomb. Some beekeepers don't like the effort of a twice-yearly harvest.

HEATHER: Nectar from heather is also a sticky, viscous product and hard to extract. Heather honey needs to be pressed out of the combs and the wax has to be recycled. However, the taste is worth all the effort.

The two medical conditions that will stop you becoming a beekeeper are anaphylactic reactions to bee stings and a chronic bad back. Though anaphylaxis is far more common with wasps, on very rare occasions bee stings can cause allergic reactions, respiratory problems and circulatory shock. A long course of immunisation will eventually desensitise the body to stings from bees or wasps, but this is only available from a few specialist centres. However the risk is very small.

The angle at which beekeepers bend over and inspect their hives is always going to put pressure on the back, both upper and lower, so any ongoing back problems could be exacerbated. Apart from the need to bend over and lever out brood frames every week, there is a certain amount of lugging and shifting of supers – the boxes used to hold the frames of honeycomb – and sometimes the whole hive.

THE HIVE

*There are several types of hive that can be used, but they all
work on the same principle. Sitting on a floor at the bottom
is a box in which the queen is located: this is called a brood
chamber and is where new bees are produced. "Brood" is the
collective name given to anything from eggs to larvae to
unemerged bees.*

A BOVE THIS IS A SERIES OF BOXES known as supers, used
for storing honey. The queen is prevented from going up
into the supers and laying eggs in the storage area by a grill
known as a queen excluder. This can take the form of a
slotted zinc sheet or a sheet of parallel wires, wide enough to
allow the worker bees through, but too narrow for the wide
abdomen of the queen to pass. The queen excluder keeps the
functions of the hive – bee rearing and honey storage – nicely
separated. The brood chamber is roughly double the size of
the super, allowing the queen a lot more space to lay eggs –
approximately 50,000 in the most popular (British) National
hive. The smaller size of the super allows bees to fill the
frames more evenly, before the beekeeper adds another one on
top. Also, the sheer weight of a honey-filled super is enough
to limit its size.

WBC VERSUS NATIONAL

Beekeepers in the UK need to choose between two types of
basic hive, the WBC, so named because it was designed by
William Broughton Carr in 1890, or the National hive. The
WBC is what most people would recognise as a beehive, with
pagoda-like stages topped by an apex roof. The National hive

looks like a series of neatly fitting wooden boxes, not unlike a packing crate with a metal lid. The difference is mostly cosmetic. The National hive is a single-walled hive that has no added protection against the elements. The WBC is effectively a slightly smaller National hive, housed within the pagoda-shaped external covers called "lifts" which come off one at a time, i.e. it is a twin-walled hive. The WBC is the only twin-walled hive on the market in the UK.

Though the WBC has the advantage of a wide, built-in landing board, the beekeeper needs to take off the roof and almost all the lifts before making their inspection, whereas with the National they simply take off the roof.

Apart from the time element, WBCs can be more disruptive to the bees as fitting the lifts brings inevitable knocks as they are put back on. This becomes quite a process when you have four supers at the end of a productive summer and have to use six or seven lifts. But the most serious drawback of a WBC hive is that it cannot be moved easily. So if you want to place the hives in an orchard for the early part

WBC hive (left) and a National hive (right)

of the year to take advantage of the apple blossom, it's more work with a WBC.

If you are planning to have just one or two hives then you will have the time to handle WBCs. They are aesthetically far more pleasing than Nationals and better at keeping out the cold and damp, which bees hate. It's recommended that hives are not kept right under trees but with a WBC you could just about get away with it. For three or more hives, using WBCs will create a lot of extra work.

One useful element is that frames (see opposite) from Nationals and WBCs are interchangeable, so beekeepers can have a WBC for show and a couple of Nationals to get the honey crop in without having incompatible frames.

Hive elements

Starting from the bottom up, a National hive is made up of a flat wooden floor with a snugly fitting – but not attached – wooden entrance block that can be removed later in the season. The block runs the length of the entrance and can be rotated to give either a total block or a narrower entrance. These are kept in place as a protection against cold, wet weather and raiding bees from other hives. It narrows the entrance and gives the worker bees, which guard the entry, a much smaller area to patrol and thus frees up other bees to do the more important job of foraging and building up colony numbers. When the narrowness of the gap is clearly creating a crush of bees as they flow in and out of the hive around May, it's time to remove it.

Above the floor sits the "brood chamber" containing what we hope is a vigorous laying queen and brood frames. On top of the brood chamber is a queen excluder, then the first super.

Above the super is a removable flat board with a hole in its centre known as a "crown board". The hole provides ventilation or feeding access and also allows it to double up as a "clearer board" later in the year by the insertion of an "escape". When the honey is harvested at the end of summer, to remove supers they first need to be cleared of bees. Beekeepers do this with a series of one-way escapes, the most common of which is the Porter Bee Escape. It allows bees back down into the brood chamber but won't let them back up into the super (see page 62).

Above the crown board is a flat roof. WBCs are identical but with dimensions that allow for one less frame in the brood chamber. While the WBC roof sits on the uppermost lift, the National roof will rest on the crown board.

FRAMES

A brood chamber and a super are simply wooden boxes with ledges at either end from which a number of frames of honeycomb are suspended. Frames have wooden top, bottom and sidebars with a sheet of foundation wax embossed with hexagonal cell-shapes to give workers a start in the honey-comb building process. When the frames hang vertically, end-on to the front wall of the hive, i.e. at right angles to the entrance, this is known as "the cold way". Turning them the other way round is known as "the warm way".

Between each super there is a 6mm gap or "beespace" allowing bees to travel laterally in the hive. Without it, once the supers are in place, there would just be a sheer wall of honeycomb stretching from the bottom to the top of the hive. Even then, when a super is placed on top of another super, the bees will do their best to fill some of the vertical gaps between the bottom bar of the upper frame and the top bar of the frame beneath.

In the UK most hives have a bottom beespace, i.e. the gap for the bees to move around is at the bottom of the frame, while in the USA they tend to use a top beespace, with the gap at the top of the box. The top beespace allows beekeepers to manoeuvre supers around a lot more easily. It only becomes a factor if you pick up secondhand supers (mostly self-constructed) where the beespace is at the top and combine them with other supers where the beespace is at the bottom.

FRAME SPACING

There are three basic systems for keeping frames evenly spaced through brood chambers and supers. The gap that the bees will leave alone is the size that they need to crawl through freely is about 6mm. It was the discovery of beespace that led to the design of the removable frame hive and modern beekeeping. There are two kinds of frame, the British standard frame and the Hoffman self-spacing frame.

The Hoffman frame has much larger wooden shoulders at the top of the frame that prevents them getting too close to each other and, even when all the frames are jammed together, there is a built-in gap for the bees to go to work. Hoffman frames are becoming increasingly popular. The alternative is to have conventional British Standard frames with spacers that slot onto the ends to keep the frames from getting too close to each other. Spacers come in two basic sizes. The smaller is suitable for frames in the brood box and the larger can be used to increase the gap between frames in supers. As the bees fill the supers with honey the frames may be spaced further apart and the bees will "draw out" the wax, making a single

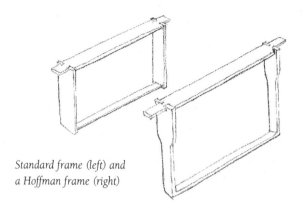

*Standard frame (left) and
a Hoffman frame (right)*

frame wider and hold more honey. When harvesting, there is
less time spent processing fewer wide frames than lots of
thinner ones.

Spacers can be metal or plastic and can come in a variety
of colours. The intention behind the colour code was so that
beekeepers could be reminded in which year the frames were
created, but this system is not used very much. However, the
colour code comes into its own when remembering in which
sequence to put the brood frames back in the brood chamber.
Because there is an element of "sculpting", when bees produce
or draw out comb, quite a few frames will curve out on one
side and curve in on the next frame along, like a curvy
corridor. Get these curves out of sequence and you can have
brood comb butting up against other brood comb and not
enough beespace.

An alternative to these systems is to use castellations built
into the ends of each super. The frames fit into pre-
determined slots at each end. This only really works in supers,
as when working with a brood box you will want to slide the
frames about as you go.

As in many things regarding beekeeping, beekeepers can never agree about frame spacing. Some think that the Hoffman frames don't give enough space to the bees, while others wouldn't use anything else. This "my way is the right way" attitude comes up quite a lot in the hobby and it's based on the belief that "if it's always worked for me, then it's got to be correct". In practice, bees are very forgiving with the people that manage them and get by despite what is done to them not because of it. There is often more than one good solution to a problem, so keep an open mind.

NUCLEUS HIVE

A nucleus, sometimes abbreviated to just "nuc", is the term used to denote a small colony of typically four frames, either in a specialist four-frame mini-hive or in a larger brood box which has had the width restricted and the entrance reduced to a couple of bee widths to prevent robbing. These have multiple uses, such as starter colonies, swarm control and for rearing queens. Nucleus hives are easy to inspect and the standard frames can be transferred to a conventional brood chamber once the colony is strong enough to start filling the extra space.

OTHER HIVES

If you buy equipment secondhand or take up the hobby in Scotland or the north of England, you may come across other hives than the National or the WBC. In Scotland, the Smith hive, which has the same internal dimensions, is popular and there are also hives such as the Langstroth, Dadant and Commercial available, though they make up a much smaller proportion of new equipment sales. The one thing that these hives have in common is that, with the exception of the Smith hive, they all have bigger brood chambers than the National.

SKEPS

The dome-shaped straw skep commonly seen on honey jars was the hive used by beekeepers of old. Before bees were kept in well-managed boxes, they were collected as swarms in straw skeps. Beekeepers left the bees to build their own matrix of honeycomb through the summer, with no careful inspections. Come the autumn, the bees were driven off with smoke or killed and the entire contents removed, pollen stores and brood cells included. It was an extremely wasteful process. Skeps are still used to collect swarms and they are the perfect shape for them to fall into, but that is about all.

EQUIPMENT

Equipment needed for extracting and processing honey is covered in the chapter on Honey Harvest on page 60.

VEIL OR SUIT

When starting in beekeeping it is important to feel safe and secure. Though many people supply bee suits, the ones marketed by specialist companies in the field tend to be the highest quality. Bees have a tendency to crawl up sleeves, under jumpers, onto ankles and up trousers, so an all-in-one bee suit and a set of wellingtons that don't gap at the calf is the perfect outfit. The better the suit, the more secure you will feel. It's better to use a veil that has cloth behind your neck rather than mesh, which over time is liable to get twisted.

GLOVES

It's best to get the highest quality gloves you can afford. However, there is a compromise to be made between the degree of security from stings and the lack of dexterity or feel that you have through your gloves. Kid leather gloves will allow you a great deal of dexterity and some protection, but they are not sting proof. Heavy leather gloves can be completely sting proof, but can be awkward to use.

It's as well to remember that when one bee stings it releases a chemical that attracts other bees to the area and makes them aggressive. Make sure you wash fabric that has been stung. Many seasoned beekeepers dispense with gloves and handle frames without protection. This not only allows for faster inspections and less disturbance, it also allows them to be gentler and more precise in their actions. However this is not recommended for the inexperienced beekeeper.

SMOKER

A good smoker is an essential part of the beekeeper's basic kit, because it's the main means of control. There is very little to distinguish between them, but the bigger the chamber, the more oxygen you can get to the combustive material, and so the less likely they are to go out. What you put inside the smoker is perhaps more important than the smoker itself.

SMOKING MATERIALS

The vital thing is to use material that doesn't go out the minute you stop working the bellows. There is nothing so dispiriting as carrying out a complicated manipulation in the hive, reaching for your smoker only to find that it no longer smokes. Bees don't appreciate the pause while you rush off to find your lighter, take your gloves off, light the wad, puff away, put your gloves back on, then return to the hive.

The goal is to produce a cool, non-acrid smoke and the following materials are all used by beekeepers.

Fuel notes

DRY WELL ROTTED WOOD Very good, but can be tricky to light if not absolutely dry

ROLLED CORRUGATED CARDBOARD Burns a bit fast but easy to light

WOOD SHAVINGS Ok but burns fast and can blow hot ash

OLD HESSIAN SACKING Some grades can be tarry and noxious; getting harder to find

OLD HESSIAN SACKING ROLLED WITH CARDBOARD Good but watch out for above

COMPRESSED COTTON WASTE Very good but you have to buy it. Lasts a long time

Hive tool

There are two main types of hive tool. Both, in effect, are scrapers and work equally well. One is the J-shaped tool; the other is a scraper tool with a curved end. It's all down to personal preference, but it's handy to have both in case one

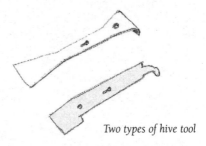

Two types of hive tool

goes missing in the long grass. Use them to scrape away and remove unwanted comb and propolis.

The tail of the "J" is used to lever up frames, the levering point being the frame next door. This is fine provided the bees haven't glued the frames so securely in place that all you do is lever the top bar off the frame leaving it sitting there with no easy means of removal!

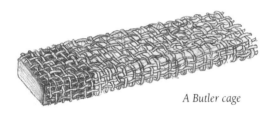

A Butler cage

QUEEN INTRODUCTION CAGE

Introducing a queen to a new colony is a delicate business. If you have invested £40 on a bought-in queen to improve the temper of your colony, then the last thing you want is regicide the moment you shut the roof of the hive. Bees can easily take against the new queen, especially if she is marked and clipped, because they might regard her as abnormal, and so the slower the introduction, the more likely it will succeed. A queen cage allows the bees to lick the queen without causing her harm. Gradually the collective pheromones called "queen substance" gets passed around and the bees begin to accept her as their own. The most basic form of cage is the Butler cage, which is jammed between the frames of the brood chamber. At one end of the wire mesh cage there is an opening that can be covered in newspaper and held in place with an elastic band. In time, the bees will chew through and release the queen. By then her new subjects will be accustomed to her and her substance. There are different forms of queen introduction cages available, including the retro-looking hair roller cage, but they all work along similar lines.

LANDING BOARD

Two good things that a WBC hive has and a National doesn't, are the landing board and an apex roof. Both can be bought as optional extras for a National hive. A landing board is not essential, but it's nice to see what the bees are bringing in to the hive – the colour of the pollen etc. Occasionally it's useful to see what they're dragging out as well. With a standard entrance, bees get airborne the instant they come into view, giving the observer little time to see what they're up to. When they have a landing board this process is more leisurely, with bees landing and walking the last few centimetres rather than hovering and waiting for a gap in the traffic. Also, in cold weather, a landing board is a much easier target for an exhausted bee to hit than a tiny space protected by a mouse guard.

MOUSE GUARD

A mouse guard is an essential item for every beekeeper to attach to the front of their hive as they go into winter. Though it may look too small, mice can easily squeeze in through the entrance space of a hive and set up home for the winter, destroying the comb and eating honey. There are a variety of devices that can be used to keep rodents out, but the most common is an aluminium strip drilled all the way across with bee-size holes. This can be either the full width of the hive or a smaller one fitted to the reduced width entrance. In warmer winter/spring weather it's always useful to watch how the bees are using the entrance with the mouse guard attached. If the holes are set too high then it's a real hurdle for the workers to manoeuvre old dead bees through the spaces.

OPEN MESH FLOORS

A great many beekeepers have replaced their traditional wood floors with open mesh floors (known as OMFs), allowing more air circulation through the hive, even in winter. This is not so unnatural when you consider that bees in the wild have evolved to overwinter in hollowed-out trees or in caves where they are protected from the top but open at the bottom. And if we use the evolutionary device of the fake forest fire in the smoker, then why not adapt their hives to nature as well?

The advantage of OMFs is that they can be used as part of an integrated pest management (IPM) system to control the

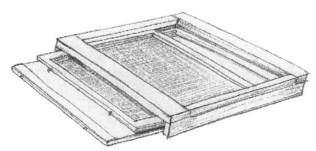

An open mesh floor for a National hive

parasitic mite varroa, which is now endemic in the UK. If the varroa mite can be encouraged to slip off the bee to which it clings – by some mechanical method such as covering the bees with talc or icing sugar – then a mesh floor will let it fall through to the ground. A solid wooden floor would allow it to climb back onto the comb or onto a passing bee again. Even without such encouragement, up to 10% of the varroa population naturally fall though the mesh and out of the colony. Also, while bees will happily propolise any mesh that is placed upon the crown board hole, they rarely propolise a mesh floor in winter.

EXTRA EQUIPMENT

A lot of the bee management techniques involve using extra hive equipment. At first this might seem a bit wearing. If you keep chickens, you don't need a couple of extra coops on standby just in case the chickens get a bit moody. In beekeeping it's always useful to have a spare brood chamber, a spare floor, a spare roof, queen excluder, crownboard, brood frames and a couple of spare supers complete with frames.

If, in your first year, you're starting from a single nucleus of four frames, and not a full brood chamber, then it's unlikely you'll need to do much management. The colony will spend the whole of the summer building numbers and building comb. By the time it comes to your second season you will know enough people at your local beekeeping association to pick up the phone and borrow what you need.

HONEY BEES

As in so many societies, it is the women that do all the work while the men laze about and think only of procreation, food and meeting up with other males. The vast majority of bees are workers (sterile females); there is a much smaller number of drones (fertile males) and one fertile female, the queen.

THE QUEEN IS THE MOTHER of all the bees in the hive and is very busy in the spring as the colony expands, laying up to 2,000 eggs a day. She lays one in each cell of the brood frame; that's about an egg every 20 seconds, which is more than her own body weight in eggs every day.

Throughout the year the composition of the hive will vary. In the winter the numbers in the hive will be as low as five or ten thousand workers, with no drones present. Come the summer, a few hundred drones will be created while the numbers of workers can rise to 40,000 dependent on the weather, the amount of forage available, the productivity of the queen and also on the size of brood chamber that she has to work with.

Bigger colonies than this are possible, but usually only with an enlarged brood chamber (some beekeepers add on a super as a kind of brood extension, known as a "brood and a half" or even put two brood boxes together called a "double brood").

QUEEN

In bee colonies there is no genetic "royal line"; the egg used to create a queen is the same as one that is used to create an ordinary worker. The difference is that the larva that will become a queen is floated to the top of the cell in which it was originally laid using Royal Jelly. This rich food is produced by worker bees from a hypopharyngeal gland on their heads, and it is the sugars in this foodstuff that make the difference between the egg becoming a sterile female worker or a sexually mature queen bee. The queen lays a normal fertilised egg into an extended "queen cell" which workers build typically at the side or on the bottom of frames. Alternatively, the bees can transfer a larva that is less than 36 hours old – ideally only about 12 – into this cell.

Unlike normal worker cells, which run parallel to the floor of the hive, a queen cell is vertically oriented. Sixteen days after the egg has been laid, the new queen emerges and she is given little respect from the workers until she returns from her first mating flight.

QUEEN SUBSTANCE

One of the driving forces that makes bees create a new queen is the lack of queen substance in the hive. Queen substance is a combination of pheromones produced by the queen and licked off her body by her adoring workers as she passes through the hive. These hormones are then passed from worker to worker in the continual transfer of nectar from mouth to mouth, giving the colony a sense of wellbeing. If the queen dies, then all of a sudden the amount of queen substance plummets and the workers rush to create a new queen using the youngest egg they can find. The same thing can happen even if the queen is alive when the hive gets overcrowded in late spring, or more typically, early summer. Sheer numbers and congestion in the hive mean that the queen substance isn't getting passed around and the concentration starts to drop. Workers are then stimulated to create new queen cells.

If a new queen mates successfully she is likely to take a significant percentage of the hive with her in a swarm. This tendency for bees to panic and create new queens is known by beekeepers as "swarminess" and, along with aggression, is one of the least desirable characteristics of a particular strain of bees.

SUPERSEDURE

A queen usually lives for around two or three years but, in exceptional circumstances, can live up to five. If she is underperforming – either as a poor layer, or a poor queen substance provider – the workers will decide to replace her. This process is known as "supersedure" and is most likely to happen towards the end of summer or in early autumn. Bees transfer their affections to the new queen, while the old one loses weight and gets fed less and less, before wasting away or getting culled.

MATING

A newly emerged virgin queen will strengthen her wings on short preliminary training flights before, weather permitting, going to "where the boys are" (called drone assembly areas) on her first official mating flight. Queens will mate with up to 20 drones on one, two or even three mating flights, until her spermatheca (sperm sac) is full of sperm. This sperm will have to last her a lifetime.

The mating act is a violent one performed on the wing. When the drone's genitalia enter the queen's vagina, they are torn off depositing their load of sperm and leaving the drone to fall away and die with his genitals still attached to his one-afternoon stand. The queen will have to clear these away before she mates again. Often queens return to the hive with some parts of her brief suitor still attached, these are known coyly as "the mating sign".

When she returns to the hive, no longer the virgin queen, workers feed her rapidly so that her abdomen swells and she becomes a large egg-laying machine. Beekeepers hope she will never leave again and often clip one of her wings to prevent her from flying out with a swarm.

The temperature for mating flights needs to be 16°C or greater, and so a colony that loses its queen between October and March stands little chance of creating a successor. Not only is it too cold, by this time in the season there are no drones about.

CHANGE OF REGIME

Though aggressive colonies of bees are never a joy to handle, there is an immediate solution to the problem. A change of queen brings about a change in the temperament of the stock, as her pheromones directly affect the hive's behaviour. A queen introduced from a gentle hive will reduce the poor behaviour of an aggressive one. Specialist queen breeders will gladly supply replacements by post but this is a very seasonal solution.

It is in the nature of docile, laid-back bees that they tend not to be as productive as more aggressive bees, but every beekeeper has to weigh up what qualities they prefer in their bees. However, as a beginner, you shouldn't tolerate or make excuses for having aggressive bees. If your bees are proving to be a nuisance, not a pleasure, then the simplest thing to do is to requeen. Killing the incumbent queen is a strange experience, because it goes against everything you practice. As a beekeeper your job is to protect the queen and worry about how well she is laying and whether the colony is "queenright".

can't stay empty for long. The emergent bees are also used to "cover" or incubate the eggs. Once laid, the temperature of the brood should stay between 32 and 36°C. This means keeping larvae warm enough in spring – by keeping bee bodies over them – and cool enough in the heat of summer – by fanning them. For this reason, beekeepers carrying out inspections of their hives in early spring need to make sure they disturb the brood frames for as little time as possible for fear of chilling the brood. After three or four days the worker bee will move on to the job of creating brood food and feeding the new larvae.

From around six days onwards, they will take off on their first training flight. Even though their jobs will still be hive-bound at this stage, bees need to get out and test their flying and navigational skills, if only for the purposes of elimination. (Only when you first become a beekeeper do you begin to understand the mysteries of the little brown stains or the thin brown smudges that appear on surfaces in the garden and most noticeably on washing hung out to dry.)

From around six to 15 days, they take on the role of wax making (see page 36) and comb building, either in the brood chamber or "upstairs" in the supers. When honeycomb in the the supers is de-capped as part of the honey extraction process, the tops of the cells are often removed along with the capping, and bees will make good what the beekeeper has sawn off. Wax is also needed to "cap off" the brood when the larvae have reached nine days old and are filling the cell.

After eight or nine days the young bees will act as nectar and pollen receptors from the foraging bees and will help pack both into their respective cells. Pollen is stored in the brood chamber, and nectar that will be used as everyday feed for the bees and larvae, known as "stores", is also placed there. The surplus of nectar will go up into the supers.

Bees guarding the entrance to the hive

Between 12 and 18 days, workers can take on a more varied maintenance role within the hive. They will remove malformed larvae from cells, eject any dead or dying workers, patch up holes or gaps with propolis and commence guard duties at the entrance to the hive.

From 18 days onwards their role is as forager, going out for pollen, nectar, water and propolis. They are no longer "nurse bees", they become "flying bees". Foraging is likely to be their principal role until they die. However, bees have a tendency to ignore their career structure when it suits them. The progression of jobs described above is only the broadest of outlines. Evidence gained from observation hives – hives that have a cut-away glass section where the bees can be observed at work – has shown that they will perform a variety of jobs through the day. And whereas bees are thought to be hyper-industrious creatures focused entirely on their work, thus giving us the phrase "busy as a bee", in reality they are far more relaxed, taking time to wander about the hive, exchange nectar with other bees, rest, lick the queen, before ambling off on another task.

A worker bee's lifespan can be as little as five weeks and as long as six months. Like cars, their lifespan is down to mileage and work rate, not age. At the height of a warm summer with long foraging days, a worker can last five weeks. From November to March, when the days are shorter and they assume a more dormant lifestyle, they will hardly go out at all. A queen will reduce her laying through the autumn and so bee numbers are at a premium through the winter months. Any bee that dies is unlikely to be replaced until spring. With the steadily warming climate and the record hot summers and mild winters, bees are becoming active for more of the winter period. If winter-flowering varieties of cherry and jasmine come into flower in January then, providing the external temperature is above 10°C, bees will go out and forage over short distances.

MAKING WAX

One of the most important jobs for the worker bee is the production of wax. Each worker has four pairs of wax glands on the underside of its abdomen from which it can secrete fluid wax, which forms as white translucent flakes. To produce wax, workers first fill themselves with honey and cluster in groups to maintain a high temperature as they metabolise the honey. The combination of high temperature and ample honey in the bee's body causes the wax glands to secrete. The wax is then collected from the pockets by the bees using their rear legs and worked by the bees' mandibles before being applied as comb.

Once their bodies have hardened, three or four days after emergence, young bees will go on what are known as "play flights". At first they will just circle the hive in gradually widening circles, and as they grow older they will fly further and spend longer in the air. It gives the bees the ability to recognise their own hive so they can locate it easily on their return. Their major navigation is done by calculating angles to the sun, and by recognising nearby landmarks, which, in aircraft terms, is the automatic pilot of the journey. The mental picture helps them with the final approaches.

GUARD DUTIES

Bees will guard the entrance to the hive if they perceive a threat. In spring when the weather is just warming up and there is little threat from wasps or robbing bees from other colonies, they might not feel the need to have permanent security. However, if the hive is knocked, then one or two bees will come out to investigate what's going on. If the source of the disturbance continues, such as a garden mower passing nearby, then they will mount a continuous guard. Apart from cold and damp, bees dislike vibration, so any mowing in front of the hives is best kept to the early morning or late evening when the numbers flying are limited.

Later in the year, when the hive has a sizeable honey reserve to protect, guards become a regular presence at the entrance. They will try to sniff any entrant to the hive to make sure they have the recognisable scent of that colony. A bee's scent is a product of the queen and the food in its gut. Because so much food gets passed around between workers, and there is regular contact by crowding and also grooming, there is a detectable colony smell to each hive. If an incoming bee doesn't possess it, then it will be challenged by the guard, which may attempt to stop it getting into the hive.

Guards seldom persist at their defence for very long. Their job isn't helped by the fact that they have seconds to carry out their task and so many of their own bees' scents are masked by the flowers they have been rolling around in. A strange bee can soon be inside and picking up the scent of the colony it enters.

The fact that guards are not great at their job is very evident in autumn to beekeepers with "glass quilts", effectively a window instead of the crown board. Take the roof off the hive at this time of year and you will see a band of workers chasing wasps round inside!

NAVIGATION AND THE WAGGLE DANCE

In addition to recognising landscape features and possessing a magnetic compass in their heads, bees are able to navigate their way back to the hive using polarised light as a guide. What's more, as the sun is gradually shifting its position in the

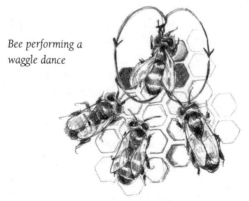

Bee performing a waggle dance

sky when they are out scrambling from flower to flower, they are able to calculate the elapsed time and work out the correct flying angle to get them home again.

In their life as foragers, if they find a plentiful supply of pollen and nectar they will pass the information on to fellow foragers back at the hive. They do this through the medium of dance.

The round dance is performed for nectar and pollen that is within 10 metres from the hive. A bee rushes around in a circle on the vertical face of the comb – first anti-clockwise and then clockwise. The interested bees that gather to watch are given a sample of the nectar, or get to smell the pollen in the bee's pollen sac. This is the only clue as to what they are looking for. Basically the dance tells them: "There is pollen/nectar close by, this is what it tastes like, go and look for it."

The famous waggle dance is more complicated and is used to communicate the position of nectar and pollen more than 100 metres from the hive. Between 10 and 100 metres, the bee uses a hybrid between the two styles of dance. A bee will dance a "squashed figure-of-eight" pattern (see illustration) on the vertical face of the honeycomb, with the position of the sun being at the top of the hive. Even though the hive is dark, bees instinctively know, through gravity, where the vertical is. The direction of the nectar/pollen they want is the repeated line in the centre of the figure of eight. So if the direction of the forage involves flying directly at the sun, the bee will perform a figure-of-eight dance with the repeated central line running vertically up the comb (6 o'clock to 12 o'clock). If the direction is directly away from the sun, then the repeated central line will be down the comb (12 o'clock to 6 o'clock). In other words, the bee is translating the angle relative to the current position of the sun to an angle from the vertical. Consider that as the sun is constantly moving across the sky, the bees' dance is therefore likewise constantly changing to represent the same forage source. If the direction of forage is at 90 degrees to the right of the sun, then the position of the repeated central line will be horizontal, running to the right of the comb etc. The famous waggle, the rapidity in which the worker performs its dance and the length of the repeated central line, are all clues as to the distance of the source. Just like the round dance, watchers of the waggle dance will also be given samples of what they can expect to collect.

DRONES

Drones are distinctly bigger than worker bees and are stingless. They are produced when the queen lays an unfertilised egg in a special (larger) drone cell. As the drone is from an unfertilised egg, he has half the chromosomes of a worker bee (16 rather than 32), a system known to biologists as parthenogenesis. Strangely this means the drone has no father, but as the queen was from a fertilised egg he does have a grandfather.

If there are no drone cells available, workers will create new ones either by ripping out chunks of normal comb or building

A standard frame with worker foundation (left) a frame with drone brood foundation at the bottom (right)

them close to the bottom bars of existing brood frames.

Because drone brood cells formed in this way often stick out into the space between the hive floor and the line of the bottom bars, and bulges sideways, when the frames are pulled upwards by the beekeeper they drag against existing cells, causing a great deal of irritation to the bees. A solution is to have one brood frame with a dedicated drone foundation at the bottom, so workers will draw out perfect drone cells in their own little mini-colony. Having all the drone brood in one place will also come in handy when trying to manage the bee mite varroa without the use of chemicals.

Each hive will produce a few hundred drones each summer, and their single task is to mate with a new queen if required. They do no work at all, instead they wait until the hottest part of the day before cruising out to "drone assembly areas" where they will hang around with other drones, awaiting the rare chance of a virgin queen turning up. As described in the above section on queen mating, if they are successful they will die the moment the mating act ends.

If they are produced in a hive where the queen is a strong layer and the bees are non-swarmy, then they will have an easy summer. However, all drones are thrown out or actively executed by the workers at the end of the summer. They will not carry passengers through the winter.

Drones in the colony are thought to increase the wellbeing of the bees. Should the queen die or stop laying for any reason and they need to supersede her, then the drones are there ready to perform their suicidal sex act. They are also used to incubate brood and, author Ron Brown believes, they act as handy inadvertent decoys for predatory birds – whereas a few large drones are expendable, the similarly-sized queen on her mating flight is not.

INSPECTIONS

The heart and soul of beekeeping is the routine inspection of the hives, carried out (temperature allowing) on a weekly basis from late April to early September. As the climate in the UK warms, it is probable that this routine will be extended into March and October. In recent years we have seen temperatures in excess of 20°C in England in October and, given the glut of ivy flowers that can provide forage at this time of year, it is as well to keep the hives monitored, even if it's on a reduced level.

THE PURPOSE OF INSPECTIONS

An inspection should give a snapshot of the health of the hive. Though the questions will vary depending on the time of the year, a beekeeper needs answers to the following:

• Have I still got a laying queen? Or, is the hive "queenright"?
• Is there pollen and nectar being stored by the bees?
• What does the queen's laying pattern look like?
• Does the brood itself look healthy?
• Are the bees crowded – do they have enough space to expand in the super?
• Are the bees planning to swarm?

Though it's nice to see just how many supers are being filled with honey, the key question is how the queen is performing. So the primary task is to see if eggs are being laid and, if so, how many. These can be quite tricky to spot near the bottom of empty cells, so the presence of young larvae is an easier gauge. Typically, the queen will lay from the centre frame of the brood chamber and move outward, only employing the outer frames when the colony is reaching its peak of production at the height of summer. A frame taken from the centre of the brood chamber should have a high throughput, normally laid with an oval of brood in the centre, with a crescent-shaped arc of pollen above. In the top left and right corners honey will be stored for nurse bees to turn into brood food. The most efficient brood-warming cluster shape is a circle. The reason it is oval, rather than a circle, is due to the slight distortion introduced by keeping bees in a rectangular box.

A BEEKEEPER'S RESPONSIBILITIES

Though many people delight in telling you that they learnt everything they know about a subject from a book, this should not be the case with beekeeping. Beekeepers have a responsibility to look after their stock and not endanger others. So it is essential that beginners watch other experienced beekeepers at work before tackling a hive themselves. How you smoke the hive, lift the supers or handle the frames are all-important contributions to the temper of

A queen cell in section

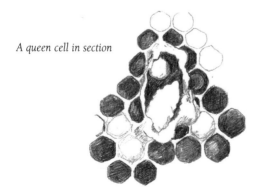

your bees. You can learn things from watching an experienced beekeeper that are impossible to learn from a book.

Beekeepers must carry out regular inspections, as they could be hosting bees that are spreading disease. If bees are regularly swarming because the beekeeper is taking no swarm preventative measures, then that is a form of neglect. There is nothing difficult about managing a bee colony, providing a beekeeper allows time in their schedule. If you are keen to spend three weeks abroad during the summer holidays you will need to arrange for someone to inspect your bees while you are away.

WHEN TO INSPECT

Inspection is a must to prevent swarms during the months of May, June and, to a lesser extent, July. During these months a planned seven-day inspection regime is required to ensure that a queen cell is not missed. At other times of the year, the inspection intervals can be further apart, dwindling to every few weeks through autumn.

The ideal time to go into your hives is the warmest part of the day. The most troublesome bees in your colony will be the older, flying bees and they will be out foraging. If the air temperature is around 12°C then you could remove the crown board to have a quick look without withdrawing a frame. You will need to wait until the temperature is 15°C or above before a relaxed full inspection can be done.

Provided it is warm enough, inspections can be made from 6am in the morning to 10pm at night. Local bee association meetings are often held on weekday summer evenings, and they will involve some communal hive inspections.

A bee smoker

SMOKING THE HIVE

The principal tool for controlling bees is the smoker. Smoke interferes with the pheromone (smell) communication "language" that the bees use to raise the "we're under attack" alarm. In addition, bees are said to have a genetic link to the time when bush fires ravaged the earth and the smell of smoke was the signal to load up on honey and move house. This is not so far-fetched when you consider that some plant species will not germinate without the presence of smoke. By puffing smoke into the hive, the beekeeper is recreating these conditions, and bees will dive into the comb and load up honey. This will allow them to go off and create new honeycomb wherever they set up house. It also makes them more docile as they are full to capacity and in no mood for a fight. The same is true when bees swarm, they too are loaded up with honey, and so the idea of an angry swarm of bees pursuing a hapless victim is far from reality.

It never seems to occur to the bees that these bush fires seem to happen on a suspiciously regular basis and their reaction is always the same.

The idea is to blow just a few wisps into the front of the hive, so that the guard bees are alerted. One of the mistakes that new beekeepers often make is smoking the hive and then starting to work on it before the message has been passed on. It's best to leave two or three minutes after smoking the entrance, so that the bees can communicate the impending forest devastation through the congested brood chamber. The other mistake is to blow too much smoke in and positively engulf the hive. This not only panics the hive, it makes the bees very angry. How much smoke you use depends on your bees, but it's another case of watching other beekeepers at work and seeing how they use smoke in the inspections before applying the techniques yourself. Once the hive has been smoked at the entrance, beekeepers should wait before removing the roof. Then it is a case of a couple of puffs in through the crown board, before going down through the supers. Quite often there will be a surge of bees upwards as you take the queen excluder off, so be ready with the smoker.

CLUSTERING

Bees cluster to generate the heat needed to incubate brood and also to survive the winter. The temperatures that bees can generate between themselves are quite extraordinary – they can also maintain themselves in the most extreme of conditions. There are tales of hives being knocked over and the bees surviving in temperatures as low as -10°C.

A swarm gathered on a branch is the perfect expression of the way bees cluster. The queen is the only one to stay at the cluster's heart while the workers take turns in the centre then move out to the fringes.

In the hive this "ball of bees" is interrupted by the vertical plains of the 11 brood frames, yet still they form an approximate ball-shape, though the effect of the frames tends to flatten that ball.

When inspecting a hive, beekeepers need to keep this 3D pattern in mind as they search through the brood chamber. One technique to get the queen to increase laying is by "spreading the brood", i.e. moving a central brood frame from the widest part of the cluster, out toward the edges. In other words, deliberately pushing out the dimensions of the ball or cluster.

INSPECTION ROUTINE

In a thorough inspection, the beekeeper will look at all the brood frames to check for signs of egg laying, to see if diseases are present, how the bees are reacting to the queen as she passes, and if the bees have been building queen cells. They need to check whether the bees are storing pollen and honey near the brood, if the brood frames are pock-marked (indicating that the workers have pulled malformed larvae from the cells before sealing them) and how much space is left in the super for expansion. This gives the beekeeper a chance to tidy up any odd bits of honeycomb that the bees have stuck in awkward places, and unglue the parts they feel ought to be stuck together – usually the queen excluder to the top of the brood chamber.

The longer a beehive is left uninspected, the greater the time the bees have to propolise everything in sight. Then the subsequent inspection becomes more traumatic for the bees, involving more jolts and bangs to loosen things.

QUICK INSPECTIONS

When you are carrying out weekly inspections, it's possible to shortcut the long-winded process of going through the entire brood chamber. Given the principle that the central brood frame is going to be the most densely laid frame and that the cluster expands in a ball-shape from the centre, then a frame two to the left will give a good indication of the colony build-up. If that, too, is almost full, then the colony is expanding rapidly. If there is no brood at all, then the beekeeper needs to investigate further.

IDENTIFYING THE QUEEN

It's one of the trickiest jobs in beekeeping – spotting an unmarked queen as she goes about her business on the frames of the brood chamber. It's important because the state of the queen can be a touchstone of what's happening in the colony. For instance, if her abdomen has slimmed right down, then she could be preparing for flight. If she is looking ragged and the workers pay her scant attention, then they may be about to supersede her.

Spotting her is akin to one of those magic eye pictures where you have to see a picture from amongst a random mass of dots. The queen can often be distinguished by the path she follows and the way the workers around her react. They will turn in towards her, to lick queen substance from her body, so the beekeeper's technique is to defocus from the mass of the bees and look for signs of unusual movement on the comb.

One great aid is to mark the queen with a dot of colour on her back. Kits can be bought which comprise of a queen cage and a set of different colours. There is a universal colouring system for each year of birth:

WHITE for years that end in a one or a six
YELLOW for years that end in a two or a seven
RED for years that end in a three or an eight

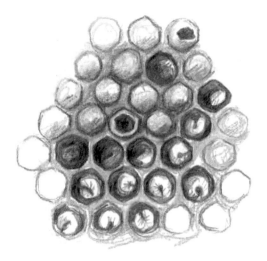

Worker larvae in various degrees of maturity

The honey harvest – not all your honey will be the same colour

The ideal location for bee hives

A bee in flight

Two bees respond to the smoker by stocking up on honey

The worker bees turn towards the queen

Drone bee

Worker bee

Queen bee

*Bees need water for metabolism and
to cool the hive in summer*

Worker bee collecting pollen

Pollinating apple blossom

Bee swarm

The inspection process should be a weekly event in May and June

GREEN for years that end in a four or a nine
BLUE for years that end in a five or a zero

This can be remembered with the phrase: "Will You Raise Good Bees?"

Anyone who buys a kit is advised to practice on a few drones before trying out the technique on a vastly more important queen.

Knowing the age of your queen is important when you have a number of hives. Alternatively, if you are buying a new queen, you can ask her to arrive marked and clipped. Some beekeepers prefer just to use one bright colour such as white or yellow as the dark colours of red and green are trickier to see, then place a coloured drawing pin on the front of the hive to indicate her age.

TIPS FOR FINDING THE QUEEN

There are other things you can do which will enhance your chances of spotting the queen.

- Use little or no smoke as this makes her run around.
- Search for her in the middle of the day whilst the bees are flying, as there will be fewer bees in the colony.
- She is most likely to be near a patch of freshly laid eggs.
- Start looking on the first comb before the brood starts and leave a space between the brood frames and the next adjacent food frame to minimise her options to hide.
- As you look at the frame think "the bee with the long legs". Keep the sun over your shoulder and watch for her running around the edge to the darker side. Queens will always head for the darkness.
- Scan methodically, breaking up clusters by gently breathing on them.

WRITTEN RECORDS

There are two kinds of records you can keep as a beekeeper. The first is an inspection log, jotting down various points you notice as you go through the hive on your weekly inspection visit. The second is a compulsory record that you will need to keep as the owner of a "food-producing animal". Legally, bees are viewed by the Veterinary Medicines Directorate as food-producing animals, and hence any medicines administered must be logged and the records kept for five years. So your records will survive much longer than your bees. It is also a good idea to keep any receipts for medicines bought, which can be used to back up your record keeping if necessary.

INSPECTION LOG

An inspection log is vital once you have more than one hive. It is very easy to convince yourself that you will remember all the facets of your two or three colonies, but as the Chinese proverb says, "the weakest writing is better than the strongest memory".

What you record on your inspection log is entirely a personal choice but beekeepers tend to have a card for each colony where they note some of the following information:

- Age of queen, whether marked or clipped
- Breed of queen
- Date that the floor was last changed
- Date of varroa inspection/varroa count and result
- Date entrance block was taken out/installed
- Number of frames with brood
- Activity on the landing board
- Date when first super added, first super filled, second super added, etc

- Health problems
- Whether drone brood visible/drones visible
- Evidence of swarming i.e. queen cells
- Maintenance completed
- Work to be done next time
- Volume of syrup fed going into winter

Some of the information will be seasonal, such as the date when autumn feeding commenced, and some of it will be weekly, for instance the number of brood frames currently covered by the bees. Estimating brood may be a broad measure, but at least it's some measure.

Some beekeepers keep a maximum/minimum thermometer in a shady spot in their apiary so they can work out the kind of temperature their bees fly at. Compared year to year, it will give an interesting picture of the climate and the bees' reaction to changing conditions – for instance, why not record the earliest time you notice the bees out foraging for crocus pollen each year?

Most importantly, your records will give you a baseline from which to judge other seasons and stop you from worrying that things are abnormal.

Colonies with Italian or Kiwi strains of bees can go into a winter cluster for a lot longer than native British black bees and local hybrids. While the British bees are out foraging in early March, the Italians and Kiwis can still be back in the hive worrying their owner that they've died out. Recording exactly when they start foraging will be a help for future seasons. Finally, if something does go wrong and your colony doesn't make it through winter, then your records may help a more experienced beekeeper to discover what you're doing wrong.

MANAGING BEES

In a perfect beekeeping season, there should be very little to do after an inspection other than to add supers and, at the end of the season, take them off again. However in certain circumstances, a little bit of bee management is called for.

MOVING HIVES

In the course of your first season you may need to move the hive. In the evening, block the entrance with a piece of foam and secure the hive by using a ratchet tie-down strap around the floor, body and crown board. If the transit time is more than 15 minutes then use a ventilated travelling board/screen in place of the crown board. Should you have a mesh floor already fitted then this is unnecessary.

A SHORT DISTANCE

Bees will always return to the same spot, and if you move the hive more than a metre they will struggle to find their own hive. If there are no other bees close by, you can stretch this closer to two metres. If you need to move your hive further, then move it in several small steps every good flying day.

A MEDIUM DISTANCE

Should you wish to move your bees less than two miles but further than you can comfortably move in short steps as above, then first you need to move the colony more than two miles away for about five to six weeks before placing the hive in its new site.

A LONG DISTANCE

If you move your bees more than two miles then the bees will realise that their surroundings have completely changed and will re-orient to their new hive position without any problem.

EXCEPTIONS

In the middle of winter, when the bees do not fly far from the hive entrance, the hive may be moved any distance without concern.

SWARMING

There is a much-quoted saying, which goes:

A swarm in May - is worth a load of hay.
A swarm in June - is worth a silver spoon.
A swarm in July - isn't worth a fly.

The logic of this is that if you can get hold of an extra hive of bees early in the year then they'll make you some honey, but by July it's too late for them to do anything. Bees will swarm as early as late April until August, but May, June and July are the key months as the weather is always warm enough for mating flights.

Breeding from swarmy bees is the most influential factor in causing bees to swarm; all other factors such as congestion, failing queens etc, are secondary.

Swarming is the bees' idea of success, and once they make up their "minds" that they are going to swarm there is little the beekeeper can do. The secret is to make them "think" that they have swarmed: this is swarm prevention. But it is better not to allow them to get into that "state of mind" in the first place: this is swarm control.

Why swarms can be dangerous

Although we've already noted that swarming bees full of honey are not naturally aggressive, a swarm can come to rest in the most inappropriate of places. I took the mildest swarm of bees away from my neighbours' front wall last summer. It had come to rest right next to the pavement and hundreds of circling bees filled the air in the road for an hour afterwards. In that situation pedestrians would have walked through the bees, got them stuck in hair or clothing and been stung.

Swarms can remain for a few days or more hanging in a bush or wherever they have settled, so that the bees start to run low on food. Hungry bees are undoubtedly more hazardous to you and the public.

So swarming can create big problems for people living nearby, and for the beekeeper it means the majority of productive flying bees leave their hive, transforming it from a powerhouse of nectar collection to a hive of nursery bees.

Taking a swarm

Collecting a swarm is not for the novice beekeeper, but it is a fairly straightforward operation. Success or failure depends largely on where the bees have congregated. The trick is to quickly sweep, drop or cut the big cluster of bees into a container such as a straw skep and leave it there on the spot, propped open at the bottom. Bees will immediately come out and wave their abdomens in the air and fan their wings. They are using their Nasonov glands, a guidance scent that other bees will pick up and use to find the entrance – the scent equivalent of an airport's runway landing lights.

Once housed with the queen, the swarm is far more

manageable. The bees can then be transferred to a brood box and ultimately moved away.

An ideal swarm collection would be to drop a cluster from a low branch straight into a brood box set up below the swarm; the roof then being promptly attached. However real life is rarely that straightforward and swarms choose bushes, walls, hedges and chimneys more often than the textbook low branch. When bees swarm they often move to a temporary staging post, while scout bees go off and look for a more suitable long-term location. For this reason beekeepers often leave "bait hives" in their garden just in case their own bees swarm.

To persuade the swarm bees to stay installed in their new home, beekeepers can add a frame of eggs or larvae, which the bees will be reluctant to leave (thinking it's their own). A taken swarm is the perfect opportunity to get new honeycomb built because swarms have a great desire to produce wax. Beekeepers will give swarms foundation to draw out, before killing the queen later in the season – and, if they prove to be disease free – uniting them with one of their weaker colonies. The queen has to go because genetically she is prone to producing swarmy bees and keeping her will just repeat the cycle.

A bee using its Nasonov gland

Once established in a hive it is important that a beekeeper monitors the new colony for diseases and the varroa mite. Only when you are certain they are not going to cause problems with your own bees should they be brought anywhere close.

BEES ARE UNLIKELY TO SWARM WHEN:
- You have a queen who is in her first year of laying.
- There is plenty of space in the hive, and supers are added in a timely fashion.
- They come from a stock that rarely swarms.
- It is August or later.
- They show few signs of building "queen cells".
- You have taken steps to prevent swarming, such as the artificial swarm method (see below).

THE ARTIFICIAL SWARM

There are many different ways of convincing your bees that they don't need to swarm, but the simplest one is to do it for them on your terms. Though the bees' insistence on flying back to within a metre of their hive can be a relocation nuisance, it can be very useful as a management tool.
- To create an artificial swarm move the original hive more than two metres away and put the entrance at 180 degrees to its current orientation.
- Place a new brood chamber and floor, plus empty frames back on the original site. The flying bees that are out foraging will come back and find a new brood chamber where the old hive used to be, but assume it is the old hive. They won't go looking for the old one even though

the population in the new building has drastically reduced. In fact, the flying bees who exit the old hive two metres away, will go back to their original location as well.

- Find the queen and take the brood frame she is on, plus the brood frame next to her and put it in the new hive (that is on the original site).
- Now on the original site you have a hive with a lot of flying bees, a queen, a few eggs and a lot of comb to build. It's exactly the situation a swarm finds itself in.
- A few metres away, you have young bees that are yet to go on their orientation flight that will start flying from the new spot. They will raise their own emergency queen cell from the plentiful eggs and larvae left over.

Ensure both hives have adequate food as their organisation is temporarily misbalanced: the original with no flying bees and lots of mouths to feed and the new hive with no stores. If part-filled supers are available then split them between the two colonies or else feed them directly with syrup (see page 70).

If it all goes according to plan, in four weeks there will be two colonies, both in a hurry to build up and not at all concerned about swarming. At the end of the summer they can be reunited (see section below), or left to form independent colonies.

UNITING

As in the artificial swarm method above, bee management often requires splitting a strong hive into two to prevent the bees from swarming. At the end of the season it may be impractical to keep both colonies going over winter, or one may have dwindled while the other flourished, or one may even have lost its queen. In such cases the most obvious option is to unite the hives.

If the two colonies were simply thrust together then the bees would fight; the trick is to allow them to blend their respective hive odours slowly, masking their individual hive smells, so that when they come face to face they don't realise that the new bees they meet are from a different hive.

- The first thing to do is manoeuvre the hives so they are within a metre of each other, side by side. That way, after the amalgamation, the flying bees won't return to their old site.

- With two queenright hives the beekeeper will have to decide which queen has to go. Experienced beekeepers shy away from sentimentalism, but it is undoubtedly a wrenching experience to do away with a perfectly healthy queen. (The alternative is to try to overwinter her in a nucleus.)

- The supers must be removed from both hives and the bees from them shaken as gently as possible into their respective brood boxes.

- Leave both queen excluders still in place.

- Put a layer of newspaper, with no gaps, on top of the weaker brood box, making a few initial tears in the paper. Then place the stronger, queenright brood box on top, so that there is now a layer of newspaper with a few tears in it between the two. As the bees take their time to chew their way through the paper, the smells of the two colonies intermingle, and they are happy to be as one.

- Shaking bees never gets them all out of a super, so when the supers are put back on top, they too will need to be separated by newspaper.

- After all the remaining brood has hatched from the bottom brood box, remove it and any fragments of paper still left between the boxes.

EMERGENCY UNITING

In certain situations you may need to unite colonies in a hurry. It may seem cruel and crude, but beekeepers either sprinkle flour or talc over both brood boxes or spray the bees with a scent such as a bathroom air freshener to unify their smell. This method also has the effect of disorientating the bees, who will go off and clean themselves and each other rather than fighting. By the end of the clean-up they will have blended together.

HONEY HARVEST

Using figures taken from Ted Hooper's authoritative Guide to Bees and Honey, *the composition of honey is, on average:*

18% water
35% glucose
40% fructose
4% other sugars, such as sucrose
3% other substances

FRUCTOSE AND GLUCOSE are the main sugars and in total honey is nearly 80% sugar. Bacteria and fungi cannot multiply in such high concentrations of sugar and so honey is easily kept for a long time. There are also minute traces of peroxide produced after the bees add an enzyme to the honey, giving it an inbuilt antibacterial quality.

However, the older it is, the more it will lose its taste. The taste element is provided by that 3% of "other substances" which are all plant derived – there are 15 organic acids, around 12 mineral elements, 17 free amino acids and up to seven different proteins. These vary according to the plants that the nectar was collected from.

In addition, plants will supply differing levels of glucose and fructose. Oilseed rape contains a much higher percentage of glucose to fructose. Honeys that are high in glucose will crystallise a lot more rapidly than honeys that have a high percentage of fructose. Honey that doesn't set so rapidly will retain its unique taste for longer and so a clear honey is regarded as having superior taste to a set honey. Though, as in all things to do with taste, it's a matter of personal preference. Some people love oilseed rape honey while others hate it.

The sugar sucrose only appears in very small quantities in British plants, though it is a major component in citrus fruit nectar.

There are many different bottling (it's done in jars, but beekeepers call it "bottling") methods or processes that beekeepers can adopt to improve the texture of their honey, including "seeding" and "creaming" but for the first season of production it is good for the beekeeper to see exactly what his bees can produce.

CLEARING THE SUPERS

Once honey has reached the correct moisture concentration in the comb, the workers will cap it with wax to keep it clean. This is the primary indicator to the beekeeper that the bees have got the water concentration down to around 18%. It can now be called honey and is ready to be removed from the hive.

The only exception is the early oilseed rape honey, which, if left until this stage, is usually mostly crystallised and not

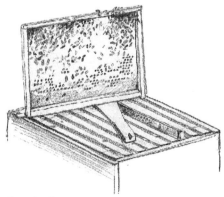

Midway through a hive inspection

extractable. To harvest OSR honey try to select a day that follows three to four days of inclement weather as this will have kept the bees indoors and they will have converted their stores to honey. Then check the combs to ensure that nectar or high water content honey cannot be shaken from the frames.

Unless you are harvesting honey from a single or monofloral source – such as apple blossom in May, oilseed rape in June, or heather in September, the main honey crop should be harvested from late August through to September.

The first thing to do is clear your supers of bees. This is done by fitting a board known as a "clearer board" with one or two "escapes" fitted in. An escape is a one-way exit – the bees can squeeze out through it, but they cannot get back in again. The most widely used is the PBE (Porter Bee Escape), but there are a number of devices that do the job including a Canadian bee escape and a radial bee escape.

Ideally you want the supers clear of bees as quickly as possible to prevent them getting agitated. Sometimes it's impossible to get them completely clear, and some beekeepers use benzaldehyde, an almond oil-smelling chemical dripped

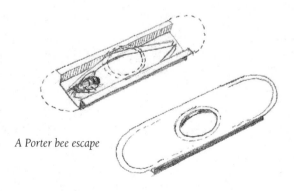

A Porter bee escape

onto a square board covered in fabric. Placed on top of the super it quickly drives the last few stragglers downwards. Some beekeepers even use benzaldehyde instead of a clearer board.

Though the entire season has involved bending at tortuous angles, the beekeeper's back can be put under even greater stress while shifting supers full of honey. Be warned. When concentrating on the job it is very easy to forget to adopt the right posture.

HONEY HANDLING

In recent years various food handling and labelling laws have been introduced to the honey bottling process. If you are producing honey for yourself, you can still eat it like a bear if you want. However, if you are selling it, or even giving it away to others, various hygiene rules apply. Equipment must be constructed from stainless steel or food-grade plastic. Premises must be provided with hot and cold water, a wash basin and a sink, and must not communicate directly with a bathroom/toilet. The space must be cleanable and have first-aid facilities. In other words, your kitchen.

The common sense rules of personal hygiene also apply, such as washing hands, covering up cuts, not smoking etc.

UNCAPPING

Supers should be stored in a "bee-proof" space and extracted as soon as possible, if only for the reason that other bees and wasps will sniff them out and try to reclaim the honey.

The thin wax capping on the honeycomb needs to be removed before the honey is "spun out". The favoured method is with a serrated knife, which is dipped into hot water to warm the blade and sliced just under the surface of the capping. The trick here is to keep the knife as hot as you can and to take off as little of the honeycomb as possible. The deeper the knife cuts, the more the workers will have to rebuild the comb in the supers when they are put back in the spring.

Alternatively, you can use an uncapping fork which has 12 or so sharp tines that prick open the honeycomb caps. This method works well on the low-lying areas of honey – natural depressions in the wax – that are difficult to slice down to, and is kind on the honeycomb walls. Unfortunately it will also produce more bits of loose wax to be removed at a later stage.

USING AN EXTRACTOR

Though frames with runny honey can simply be left to pour out under gravity into a collecting dish, the most efficient form of extraction is in a tangential extractor that is really a centrifuge in a big plastic or steel tub. The frames are uncapped and put into a wire cage – commonly four at a time

A centrifugal honey extractor

to balance each other– which can be revolved at speed. The centrifugal effect flings the honey onto the extractor wall where it slides down and is channelled via a sloping bottom into a sump with a tap. The trick is not to spin too fast or the weight of the honey on the inside of the frame, pushes against the now-empty extracted side of the frame and breaks it. Frames are then turned round and the other sides are spun out, though this time there's less need for caution because there's nothing pushing on the other side. In a radial extractor the frames are placed like spokes on a wheel and both sides of the frame are then extracted simultaneously. Neither extractor is suitable for thixotropic honeys like heather. The combs need to be crushed to extract such honeys, not recommended for the beginner.

Honey extractors and the accompanying storage buckets and tanks that are needed are big bits of kit to buy in your first year as a beekeeper. A lot of beekeeping associations have communal equipment that they lend out and, as you can probably guess, the premium time is September/October. Alternatively, if you get friendly with a beekeeper they might lend you theirs – most stand idle for 363 days of the year.

It's probably a wiser investment to buy buckets or tanks for your honey, because then you are not rushed into bottling it straight away, which is not always a good idea especially with fast-setting honeys such as oilseed rape.

FILTERING THROUGH

Before you can bottle your honey you need to filter it. The extractor will spin a fair few bits of wax into the honey and that, along with other odd bits, needs to be taken out. Over time all wax will float to the surface, but because the honey is viscous that process can take weeks.

Honey is much easier to filter and wax settles out faster when it's warm and many beekeepers convert a large old fridge into a "warming cabinet". Because fridges are well-insulated, a single 40W light bulb in the bottom, ideally with a thermostat, is all it needs to generate reasonable temperatures.

Honey can be warmed but it must not be overheated or it changes the taste. The warming process will thin the honey, allowing more of the wax to float to the surface quickly where it can be skimmed off before filtering. It will also go through sieves quicker.

Bottling experts recommend that you use two or three elements together, such as a coarse kitchen sieve, followed by a fine kitchen sieve and finally a nylon filter cloth. If you use a finer sieve, the process will be faster if the honey is gently warmed first. The honey should flow from a large container fitted with a honey tap so that you can quickly turn the flow off if one of the sieves is getting overwhelmed. Equipment suppliers produce special honey tanks in varying sizes but Graham and Annie Law at www.beeginners.info recommend five-gallon home-brew fermentation bins fitted with honey taps; though they advise that these can get quite heavy when full.

BOTTLING

There are a great many honey jars available; square ones, skep-shaped ones, tall ones, but the standard 454gm 1lb round jar will always be the cheapest to buy. Choosing the jar is the easy bit; making sure that your label contains the correct legal information requires more thought. Food handling and labelling standards may change, so it's best to check on the Food Standards Agency website, where downloads that relate to honey bottling are available.

In 2016 beekeepers need to put the following information on their labels.

- That it is honey.
- Its net weight in grams. If you include pounds, the metric measure must come first. The font size of the lettering should be 4mm or larger.
- The name and address of the producer or seller.
- The country of origin.
- Either a batch number or a "Best Before" date. Honey will easily keep for two years or more and you can specify: "Best before 2018". However if you don't use an exact date, you will have to put on a batch number as well. So it's easier to put: "Best before 31.12.2017" instead.
- This information needs to be within what is described as the "Field of View" i.e. can be seen when looking at the main label. Beekeepers tend to opt for a grand label on the side of a jar, with the above information on a sticker on the lid.

Optional information

- You can describe what region your honey comes from on the label such as: Sussex Honey or Weald Honey.
- You can describe it as a particular kind of honey providing it has 75% of that honey in it. Examples are oilseed rape honey that is distinctively white in colour or heather honey that is dense and granular and is sold at a premium.
- A tamper-resistant seal is a good safeguard but not a legal requirement as yet.

Honey, I wrecked the kitchen

The honey extraction and bottling process is a sticky and time-consuming business. Supers and frames will often have highly staining globs of propolis attached to them that can get transferred to kitchen surfaces. The sheets and sheets of discarded cappings will need to be put somewhere to be drained of their excess honey, while the serrated knife will need to be constantly wiped clean of wax. Some beekeepers are threatened with divorce directly after extraction has taken place, so it's best to choose a time when the whole area can be dedicated to the task for a length of time.

Preparing for winter

Once the honey is removed from the hive, beekeepers are effectively managing the hive into winter. This involves treating the bees for varroa, giving them the reserves to survive till spring and battening down the hatches. This is really the embryonic start to next season, as you now decide how many and what colonies are fit to survive the winter by merging weaker colonies into stronger ones (never strong into weak).

Feeding

Bees store honey to feed themselves through the winter months. They also store honey as part of the natural biological drive to expand and reproduce: in equatorial regions where there is very little seasonal change, bees can forage all year round, so the honey they collect must be fuel for swarming and reproduction.

At the end of August, when a season's work is done, the European bees' honey is suddenly robbed from them by the beekeeper. It is then they must be fed to make up the reserves that will take them through the winter. Thankfully bees are quite happy to receive sugar solution in exchange for their carefully collected honey; they don't even have to go outside the hive to get it. Indeed, refined sugar is a better source of energy than the hit and miss honey crop where the early OSR would now be rock hard and unusable by the bees in the depth of winter when they cannot forage for water to dilute it. Bees need the "stores" for their own needs and also to provide heat for the cluster. In the spring, stores are used by the workers to produce "brood food" fed to larvae once the queen starts laying.

Bulk and contact feeders

There are two main kinds of feeder; bulk feeders and contact feeders. A bulk feeder is one where the bee travels up inside the feeder and sucks directly from a lake of syrup.

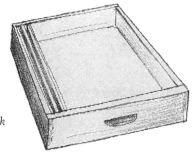

Ashforth bulk feeder

These are good devices for giving bees a lot of sugar quickly, examples of which are the Ashforth, Adams and Miller feeders. They are all around the size of a super and are placed above the bees directly on top of the brood chamber.

The alternative is the contact feeder that looks like a small plastic bucket with a lid.

In the centre of the lid is a fine gauze through which bees are able to suck sugar syrup. The feeder is filled with syrup and inverted till it stops dripping, then placed on top of the crown board with the gauze lined up over the hole. As the bees empty it, the vacuum created at the top of the container keeps it from dripping out. Beekeepers will need to add an empty super on top, so that the roof rests on this and not on top of the plastic contact feeder.

Contact feeders are generally used as top-up feeders, or as stimulus feeders in the spring. A surge of sugar in the hive from a contact feeder will get the queen to start laying rapidly as though she were in the midst of a honey flow. She's not to know it's Tate and Lyle's finest!

Contact feeders can also be used if the bees are short of stores – for whatever reason – in the middle of the year. Part of the weekly inspection is a quick visual check to see that the bees have enough stores in place.

How much syrup?

To make syrup, 1kg of sugar should be mixed with a litre of water (known as 1:1) and heated till the crystals dissolve. This is a good average consistency to feed to bees, but only once cooled down.

A strong colony in a National hive brood box will need around 16–18kg (35–40lb) of stores to keep them going through the winter. The bees will pick up some of this themselves in September and October, but will probably need around 10 litres of syrup providing around 8kg of stores.

When the bees pack the sugar solution into cells they need to get rid of the excess water. In the past, beekeepers have been advised not to feed later than mid-September because cold damp weather makes it difficult for the bees to do this.

However, now that parts of the UK are experiencing 20°C temperatures past mid-October, it's advice that may need to be revised. What is right for a Cornish beekeeper won't necessarily suit someone in Perthshire. So listen to local beekeepers because they will have the most relevant advice.

EMERGENCY FEEDING

It's very difficult to know what is happening to the bees once they go into their winter cluster and the only way to know if the hive is running low on stores is to weigh or "heft" it. One emergency method to get sugar to the bees is by placing a block of candy on top of the crown board or even a bag of sugar split open with a cup of water added.

Never feed them honey unless it's their own. Other honey, especially foreign honey, may contain diseases or spores of American Foul Brood.

A quick way to make candy is to take 2kg sugar and drop it into 0.5 litre of water and heat it to 112°C. A teaspoon of cream of tartar helps to stop it setting too hard.

Mix the sugar into the water slowly and bring to the boil for 10 minutes stirring occasionally. Test the mixture by dripping onto a cold plate: when it forms a soft solid mass, it's ready. Take off the heat and stand the saucepan in cold water for about 20 minutes until you see white streaks appearing.

Then stir vigorously and pour into 2kg containers such as old ice cream tubs and leave to set.

There are other ways of emergency feeding: instead of making candy, you can buy baker's fondant from catering suppliers, or just pour slightly damp sugar on top of the cover board.

THE VENTILATION DEBATE

A great way to start an argument with a beekeeper is to find out if they're pro-ventilation or anti-ventilation in winter and tell them they're wrong. Within the craft there are very strong advocates for leaving the hole in the crown board open in winter to allow a through-flow of air in the hive, or alternatively raising the crown board up by a few millimetres to allow the damp air out at the sides.

There are those who would argue strongly against it, though. They point to the fact that bees tend to seal up any cracks that they find in the hive and if a mesh is placed over the hole in the crown board then the bees will propolise it. And the bees must know what they're doing.

With no definitive guidance one way or the other, find the way that works for you and stick to it. If you have warm, wet winters then it's probably better to ventilate and if you have cold, dry winters then it's probably better not to.

PROBLEMS

There are a number of pests and diseases that affect bees, ranging from minor irritants to the colony to those that threaten its existence. Mice are a nuisance in winter but can be guarded against; wasps can be repelled providing the bees have a small enough entrance to guard, while wax moth is down to the beekeeper to spot and eradicate. The most critical time to get rid of wax moth is going into winter when the hive will be undisturbed for many months and the wax moth larvae can plough through many frames that the bees have deserted.

VARROA

The most crucial and widespread debilitator of bees is the mite *Varroa destructor*, previously known as *Varroa jacobsoni*. The female mite is around 1.7mm long and crab-shaped, with eight legs to the front and a wide, oval-shaped body. They are difficult to see when attached to the bees because they grab on to the underside of the bee's abdomen. Varroa mites feed by piercing the membrane between the plates of the abdominal

Varroa mite

segments and sucking the bee's blood. Once established in a colony they will breed quickly and are near impossible to eradicate entirely.

The presence of varroa is not terminal to a colony, unless the numbers reach infestation levels. Bees can still go about their business with a certain number in the colony, but the beekeeper has to monitor the levels of varroa in case they start to get out of hand. Because no chemical treatment should be applied while there are supers of honey in place, beekeepers will need to use the anti-varroa medication in the spring and winter (though if the level of mites exceeds 2,500, they must be treated there and then). At the time of writing, the official advice is to try to keep the mite population below 1,000 to minimise damage to the colony.

Beekeepers can very roughly estimate the level of infestation by counting the number of mites that die in a seven day period and multiplying that number by 30 between May and August, or by 100 at any other time of the year.

To do this, a special sticky floor insert has to be added which will trap the dead mites and stop them blowing away or, alternatively, a special varroa floor can be used which has a tray underneath a mesh bottom. The varroa mites can be taken from a piece of paper or stiff card placed on top of the tray.

Throughout the season there are mechanical methods that can be applied and these in conjunction with various chemical treatments, form the basis of what is termed an integrated pest management (IPM) regime.

VARROA REMOVAL

There are a number of different ways that varroa can be removed from the colony but the key methods are:

HARD CHEMICALS Apistan/Bayvoral strips: pyrethroid-based pesticides, once billed as the leading treatment for varroa. The chemical is released from a slow-release plastic strip that is dangled between brood frames near the centre of the brood chamber after the honey has been removed.

It must be said that the pyrethoroid-based chemicals that have been very effective are now losing their potency as the mite has evolved to be resistant to this family of pesticides. It is therefore important to keep up to date on the latest methods and treatment for this pest in your area of the country.

Oxalic acid: this can be used in the winter. The hive is opened and 5ml of the premixed solution is trickled over each seam of bees.

SOFT CHEMICALS Thymol: thymol is known as a "soft chemical" because it occurs naturally, though not in hives. Thymol can be added to sugar syrup as an anti-fungal and anti-fermentation agent, but most importantly thymol crystals can be used to treat against varroa. The crystals need to be placed into a specially designed thymol frame that has a compartment to hold the crystals, underneath which normal foundation wax can be added. It is also available as a gel in a tray that is placed on top of the brood chamber.

Formic Acid: formic acid pads, sold under the proprietary name Mite Away Quick Strips (MAQS) are a very recent and convenient treatment. Two starchy gel pads are laid straight onto the top bars of the brood chamber, and left for a week. The formic acid vapour kills or stuns the varroa and they fall off the bees and into the bottom of the hive. Unlike thymol treatments this can be used with honey on the hive as it is flavourless and is naturally present in honey anyway.

MECHANICAL Drone brood: varroa prefer the larger cells of drone brood to lay their eggs in before they are sealed over. If specific drone foundation is used in the brood chamber then once the brood is capped the unemerged drones are sacrificed, along with a disproportionate number of varroa mites. There is also an entrapment method where the queen is restricted to laying in one particular frame using a special queen containment cage. Because varroa are drawn to lay their eggs in uncapped brood cells, and these will soon become the only suitable cells in the hive, again a disproportionate number of varroa mites can be killed by sacrificing this one frame. Both methods waste colony resources, but are a good non-chemical solution to a persistent problem.

OTHER PESTS

ASIAN HORNETS have been found in France, and could potentially cross the Channel. Hornets hunt bees by "hawking" around the hive entrance, and seizing clumsy bees returning to the hive laden with nectar.

A single hornet can take 50 bees a day; a nearby nest can wipe out even a strong hive very quickly.

SMALL HIVE BEETLES have been found in southern Italy, and could potentially spread through Europe. Native to sub-Saharan Africa, the larvae tunnel through comb, honey and pollen stores, spoiling them, and a heavy infestation can cause the bees to abscond.

Bee diseases

Just as humans can contract many diseases so can honey bees. Bacteria are the cause of the two most serious in bees, American Foul Brood (AFB) and European Foul Brood (EFB), both of which must be notified to the National Bee Unit if discovered. The local bee inspector will deal with any outbreak, often by destroying the bees.

EUROPEAN FOUL BROOD creates yellowish larvae that lie in twisted positions; bees find it easy to spot these malformations and eject the larvae themselves.

The National Bee Unit publishes a downloadable guide to foul brood identification and control.

Recent research has shown that EFB can be effectively treated by a technique known as a "shook swarm" and there is some consideration being given to removing EFB from the class of notifiable diseases. As before, it is important for the beekeeper to stay up to date.

AMERICAN FOUL BROOD is potentially more serious because it produces dry spores that can stay in the frames for many years to come and is a potential source of re-infection.

With AFB, the brood pattern is scattered with the wax cappings sunken and perforated. There is often a black scale that the bees find difficult to remove.

Due to the long latency of AFB, buying second-hand hive equipment can be a risky business and such equipment must be thoroughly sterilised by using a flame from a blow torch to scorch all internal surfaces, especially corners and cracks.

NOSEMA is a micro-organism that a lot of bees carry which can cause dysentery when the bees are stressed.

CHALK BROOD leaves mummified, chalky larvae in the brood frames and is not considered serious.

SAC BROOD is not a serious disease, but can be worrying when first encountered. It is a viral disease, that causes the bees to die while in the sealed brood stage. The larva swells up with fluid, and it dies on its back with its head turned up in the entrance of the cell. The upturned point of the head looks like an old-fashioned Chinese slipper. There is currently no treatment. In severe cases the beekeeper might consider requeening with an unrelated queen.

VIRUSES: As well as sac brood, bees are afflicted with an array of other viruses including Deformed Wing Virus, K-Wing Virus, and Chronic, Acute, and Israeli Bee Paralysis Viruses. They are difficult to treat and are perhaps best seen as an indication of other pest levels. Varroa and other mites are vectors for many of these diseases, so if symptoms are noticed, it may be time to consider treatment/requeening.

COLONY COLLAPSE DISORDER: The causes of this are still relatively unknown. What is known is that sometimes an otherwise healthy colony that is thriving one week can the next week be reduced to only a handful of bees. The colony has not absconded, as the queen is still there with plenty of food stocks and brood to raise, but her workforce is much reduced. It was particularly noticed first in the US, where there is a high proportion of beekeepers hauling hives around the country for pollination purposes.

It appears that there is no single reason for this, but it is suspected that a variety of factors is at play, weakening the colony. Stressors on the colony include bee diseases, pesticides and herbicides, monocrop agriculture and interfering beekeepers.

In short, no-one knows, but it definitely exists. Treatment is preventative, as once the colony has collapsed there is no helping it. Obviously many of these stressors are outside our control, but we can focus on keeping our colonies relatively varroa and nosema free, and disturbing them only for essential inspections.

FRIENDS' "SWARMS"

Once people know you are a beekeeper then you will be asked a great many questions about all kinds of bee-like flying insects. People will tell you they have a swarm of bees when they have: large bumble bees flying into a compost heap; wasps in a roof space; and solitary bees boring into the loose pointing around old brickwork.

Before you jump the gun and go out to investigate their "swarm" ask them how big the bees are, if they're carrying pollen in on their back legs or if they're coming out of a wall.

Most members of the public cannot differentiate between the honeybee and wasps, bumble-bees or solitary bees.

There is a series of questions you can ask:

Q1 *What do they look like?*

Yellow and black = wasps

Fat and furry = bumble bees

Q2 *Are they concentrated around a wall or lawn?*

This is usually a trait of solitary bees. Solitary individuals tend to choose the same location or time to emerge.

Q3 What are they doing?

A swarm in a cluster = honey bees. Establish its location as this will affect how you might deal with it.

The advice you give now depends on what you have ascertained from the above...

BUMBLE BEES & SOLITARY BEES: Try to persuade friends to live with them as they are non-aggressive and under threat. They will only be there for that year and their nest will be abandoned for the winter and not used again. Animals and children should be kept from interfering with the entrance but otherwise they will make good neighbours. Attempting to move a bumble bee nest usually results in its demise.

WASPS: You should advise them to contact a specialist pest control company. If the nest is accessible, there are sprays available from shops that can by sprayed on to the outside of the nest, killing the insects inside. Specialist companies use sprays and chemicals that are not available to the general public and are more effective in difficult locations. With modern litigation it is inadvisable to attempt to remove a wasp nest yourself.

HONEYBEES: You are a beekeeper and every beekeeper has lost a swarm or two, so you should try to help where possible. If it is impractical for you to attend in person then try to get a local beekeeper (whose bees they may well be) to help. Your local association may have a swarm coordinator or provide you with a list of members. If you attend in person do not put yourself at risk, swarms already in a chimney or an inaccessible void cannot be dealt with by an inexperienced beekeeper, so take advice.

THE BEEKEEPER'S YEAR

*Though the British Isles are cosseted by the warming effect
of North Atlantic drift, there are significant differences in
climate experienced by beekeepers in Devon and those in the
north of Scotland. The following month-by-month guide is
an average.*

JANUARY

- Check hives for wind or storm damage.
- Check mouse guard is still in place.
- Though snow is less and less part of our winters now, bees can get confused by it and break out of the cluster to go and see why there is so much reflected light outside. So remove any snow from the front of the hive and also stop it blocking the entrance.
- Make sure branches haven't fallen against the hives and that the entrance is clear.
- Take a look underneath the roof to ensure nothing is hibernating in there – especially in WBCs.
- Look for signs of flight on warmer days – bees will be keen to defecate the minute they get outside the hive.

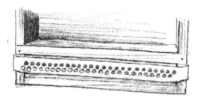

A mouse guard

Signs of activity are far more evident if you have a landing board or a clear area in front of the hive.

- Find out from your local or county association if there is a programme of winter lectures. The new year is often when most bee talks happen.

FEBRUARY

- Check hives for wind or storm damage.
- Look for signs of activity: with crocuses now in bloom bees will start to forage on warmer days.
- The early pollen stimulates brood rearing and food consumption rises dramatically. Therefore food within the hive may be running low and the bees might need feeding – this is a difficult judgement to make without looking inside the hive, so it's probably better to be safe than sorry. At this time of year it is best not to feed syrup, as the high moisture content can cause excessive condensation within the hive. Use candy, see page 71.
- Make repairs to equipment, update your records, order in new supplies, make up frames with foundation, assemble new supers etc, in readiness for the coming year.

MARCH

- Though nectar from early cherry and fruit trees should start to come in, keep an eye on the weather. A warm spell followed by a cold snap may keep the bees inside.
- Brood rearing is ramping up, making this the most dangerous month of the year, as there are an increasing number of mouths to feed with rapidly diminishing stores and old bees flying in uncertain weather. Keep a close eye

on the stores. It's ok to open the hive briefly to check stores on mild days without removing frames. Feed syrup as necessary.

- On really warm days carry out your first, brief inspection. Quickly check how many bees you have, whether you have eggs or larvae i.e. is the colony queenright, and if there is enough food, then close up again.

- Foraging should begin in earnest in warm weather. British black bees will be out earliest, while colonies with Italian/Kiwi queens may take longer to get going.

- Change the floor of the hive (see page 74). This will help you get rid of all the detritus that has fallen there through the winter months and monitor approximate levels of varroa mites.

- Before the honey flow starts it is possible to dose the hive with a varroa medication such as Apistan or Bayvarol. MAQS strips can be applied if the temperature is above 10°C; see page 75 for other varroa advice.

APRIL

- Make your first full inspection on the warmest (15°C+) day possible. This will help reduce any chance of chilling the brood. Use the opportunity to do some spring cleaning in the hive, removing odd bits of comb, etc.

- Check to see if any of the old honeycomb needs replacing. Brood comb darkens in colour as it ages. This is the time to swap old frames for new. On average, plan to renew around 25% of your brood comb every year. The wooden frames can be cleaned up for reuse.

- The mouse guard can be removed but the entrance block should still be in place.

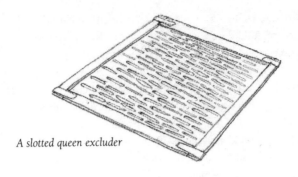

A slotted queen excluder

- If you have a spare queen excluder swap it over, as the bees will have propolised the old one to minimise draughts.
- In areas with a high density of cherry, apple, hawthorn and pear blossom, beekeepers should look to put supers on early. If an apiary is right next to an orchard, it's not unknown for a single National hive to fill two supers of honey.

MAY

- Regular weekly inspections should now be possible as the colony starts to expand rapidly.
- Swarming becomes a strong possibility from May onwards and so attention must be paid to the behaviour of the bees and, if necessary, preventative measures taken. In particular watch out for queen cells being built which gives you an early warning of an impending swarm.
- Check to see if an oilseed rape crop is within your bees' three-mile foraging radius. OSR honey crystallises rapidly and then cannot be removed by the bees, so it needs to be harvested early.

- Supers need to be added as and when necessary. Keep an empty super on top through to July; this extra space helps lessen the chances of a swarm.
- With the entrance becoming a high-traffic area it might be time to remove the entrance block and allow access across the full entrance space.
- Monitor levels of varroa through a mesh floor or using a floor insert.
- Make sure grass doesn't grow up in front of the hive and interfere with the bees' landing area.

JUNE

- Continue regular weekly inspections for signs of swarming, colony health and colony progress.
- Monitor for signs of robbing from weaker colonies and, if necessary, re-introduce the entrance block.
- If the temper of one of your hives is not improving now is the time to requeen.
- You can decide to split your powerful colonies to increase stock. This will give you more bee management options later in the season. Though it won't net you so much honey, it may prevent a potential swarm.
- Oilseed rape honey will need to be removed by this point in the season before it sets solid in the comb.
- One of the mechanical methods of reducing the varroa population is by removing whole frames of drone cells all at once.
- Beware the "June gap". In some parts of the countryside there is a lull between the dandelion and blossom crops of spring and the big tree flowers such as lime trees in July,

and bees may even need feeding. This shouldn't be a worry if your bees are visiting domestic gardens, as there is a wealth of forage to be had.

JULY

- Continue regular weekly inspections for signs of swarming, colony health and colony progress.
- Add supers when necessary – check to see that the bees are capping off honey in the lower supers. You can now force the bees to fill the supers before adding an empty one as the swarming urge is lessening.
- Ensure that the bees have sufficient ventilation – a wide entrance and/or mesh floor is fine.

AUGUST

- Swarming is much less likely now and supersedure is possible. If you have a two-year-old queen, the bees may replace her. Look out for one large queen cell in the centre of a comb; consider breeding from such bees if their temper is also good. If the bees don't replace the queen, you will have to step in and do it yourself whilst the weather is warm enough for mating flights and there are still drones about.
- Usually the end of August is the time when beekeepers remove honey from the hive, but if the weather is good, the forage plentiful and the bees are out collecting then it can wait a few weeks longer.
- Order your honey jars and plan the honey harvest.

- Remove supers from the hive using a clearer board.
- Though few beekeepers have a means of weighing their hives, it's a good reference point to see how heavy your hive feels with supers off, but roof on. This is known as "hefting" the hive. It will allow you to see how much weight the bees put on after they are fed, and how light the hives becomes in the spring. Beginners are often not good at this, especially if your empty hives are different styles and weights, but practice makes perfect.
- Begin feeding sugar syrup to replace the "stolen" honey. Whilst the weather is good this can be from a bulk feeder such as an Ashforth or Miller feeder.
- Fit the entrance block back on. If you have a problem with wasps think about reducing this space even further to help the bees defend their home – wasps seem to fly at lower temperatures than bees and get in the hive early in the day.
- Once the honey crop is removed it is possible to treat colonies with varroa medication if necessary.
- Assess which colonies have the strength to make it through winter and unite weaker colonies into strong ones to give them a better chance.
- Extract the honey. You can put the just-extracted "wet" supers back onto your strongest colonies as they can defend themselves against robbers who will be attracted by the smell. If you put a crown board with an open feeder hole between the hive and the wet supers the bees will move the remnants of the honey back into the brood chamber. When the bees have cleaned up the empty supers – usually in a few days – the bees can be cleared down and the dry supers stored for next year.

- Watch out for wax moths in the hive, or in stored honeycomb. In the summer, when the bees are active, the moths will lurk in the far recesses of the hive. The minute the bees start to cluster in cold weather they will take advantage and move onto comb.
- Feeding should be completed by now. When the weather is colder bees prefer contact feeders.
- Make your final inspection of the season before closing up the hive for winter; check that the bees have sufficient "stores".
- Check that the fit of the hive is good so that the bees don't have to waste time and energy propolising any gaps.
- Fit a mouse guard in front of the entrance block, but keep an eye on it. If the colony is still quite active it can be a real hurdle for bees trying to remove the sick and the dead.
- Oxalic acid treatment can be applied.

Jar of home-produced honey

- Feeding should have finished by now.
- Add an insulating quilt on top of the crown board, these can be bought specially or made from offcuts of carpet. Some beekeepers prefer to keep a square of carpet tile all year round and swear by the method.
- Ensure the roof is weighted down and that branches aren't dripping water onto it.
- If the hive is in an exposed location or there is a risk of vandalism using a ratchet tie down strap around the floor and crown board gives additional security and peace of mind.

DECEMBER

- Check for storm damage.
- Oxalic acid treatments can be applied.
- Go to your local association winter meetings.
- Plan for next year.
- Time for some PR duties. Give a jar of honey to all your neighbours and say, "There you are, that's honey from your garden!"

RESOURCES

YOUR LOCAL ASSOCIATION

There is nothing to match the fund of knowledge that will become available by joining your local beekeeper's association. To find your local association look at the relevant websites below which carry up-to-date contact information on each affiliated local association.

British Beekeeping Association
www.bbka.org.uk
The BBKA has over 60 different county or area divisions. Some counties will have a single association, while others, such as Yorkshire, have over 20 local beekeeping associations.

Scottish Beekeepers Association
www.scottishbeekeepers.org.uk

Welsh Beekeepers Association
www.wbka.com

Ulster Beekeepers Association
www.ubka.org

National Bee Unit
www.nationalbeeunit.com

BOOKS

GUIDE TO BEES & HONEY
by Ted Hooper
If you are only going to buy one other book on beekeeping, this is it.

BEES AT THE BOTTOM OF THE GARDEN by Alan Campion
This is one of the best beekeeping books for beginners, written by a real hands-on beekeeper.

BEEKEEPING: A SEASONAL GUIDE by Ron Brown. A classic guide to beekeeping throughout the year. Helpfully organised month-by-month.

BOOKSELLERS

Northern Bee Books
www.northernbeebooks.co.uk
01422 882751

PUBLICATIONS

BEECRAFT
www.bee-craft.com
email: editor@bee-craft.com
01733 771221 (subscriptions)
The official publication of the British Beekeeping Association.

BEEKEEPER'S QUARTERLY
www.bkqonline.co.uk
01422 882751
A broader view of the craft with news and features about beekeeping from around the world.

EQUIPMENT SUPPLIERS & MANUFACTURERS

E.H.Thorne
Beehive Business Park
Rand
Nr Wragby
Market Rasen
Lincolnshire LN8 5NJ
www.thorne.co.uk
01673 858555

Thornes are the biggest and most comprehensive bee equipment suppliers in the UK with branches in Wragby, Tayport and Windsor.

BEEKEEPING SUPPLIES

National Bee Supplies
Hameldown Road
Exeter Road Industrial Estate
Okehampton
Devon EX20 1UB
www.beekeeping.co.uk
01837 54084

CLOTHING

B.J. Sherriff
Carclew Road
Mylor Downs
Falmouth
Cornwall TR11 5UN
www.bjsherriff.co.uk
01872 863304

BBwear
Unit NP1
Rosedene Farm
Truro
Cornwall TR4 9AN
www.bbwear.co.uk
01872 562731

INFORMATION WEBSITES

www.beeginners.info
Almost every question a novice beekeeper asks is answered here on Graham and Annie Law's website.

www.dave-cushman.net
For those seeking more in-depth information about beekeeping in the UK Dave Cushman's website covers a diverse range of subjects.

GLOSSARY

APIARY A collection of two or more hives with colonies of bees present

BEESPACE The space needed for bees to do their work in between frames

BROOD The larval form of bees, from eggs through to the capped-off pupating form

BROOD CHAMBER The lower bee chamber where young bees are raised

CLEARER BOARD The board used to clear bees from supers prior to the honey harvest: it will contain one or more bee escapes

CROWN BOARD The board placed on the uppermost super, with a hole for ventilation

DRONE The stingless male bee

FLIGHTBOARD The wooden landing strip in front of the hive

FOULBROOD A serious bee disease which comes in the form of American Foulbrood (AFB) or European Foulbrood (EFB)

FOUNDATION A sheet of wax on which is embossed a hexagonal pattern, a starting point for bees to build out comb

GUARD A bee tasked with monitoring who comes into the hive

HOFFMAN FRAME A self-spacing frame

LANGSTROTH HIVE The most commonly used hive in the world

NASANOV GLAND A gland in the worker bees' abdomen that is used to attract bees back to the hive

NATIONAL HIVE A square, single-walled hive, the most commonly used in the UK

NURSE BEES Any worker that is involved in rearing brood

OMF Stands for Open Mesh Floor, increasingly popular as a means of combating varroa

OSR The bright yellow oilseed rape crop that produces nectar which will crystallise hard in the honeycomb

QUEEN EXCLUDER A grill that the queen cannot get through, separating the brood chamber from the first super, thus preventing egg laying in the honey storage area

QUEEN SUBSTANCE A mixture of different pheromones licked off the passing queen by workers

PORTER BEE ESCAPE A one-way exit for bees, used in clearer boards

PROPOLIS Collected from the bark and buds of trees; bees use it to block up holes and glue things together in the hive

ROYAL JELLY The sugar enriched brood food given to larvae in queen cells which helps them develop into queens rather than workers

STORES Honey stored around the brood chamber for feeding the larvae

SUPER The upper chamber(s) where the honey is stored

SUPERSEDURE The bees' own method of replacing their queen without the need for swarming

SWARMING The bees' attempt to split the colony in half and find a new home in the process

VARROA A debilitating blood-sucking mite that infests bees

WAGGLE DANCE The figure-of-eight dance bees use to communicate good places to forage

WBC HIVE The traditional-looking "ye olde" hive

WORKER A sterile female bee

INDEX

94